EINSTEIN ATOMIZED

SIDNEY HARRIS

EINSTEIN ATOMIZED

more ˄ SCIENCE CARTOONS

C

COPERNICUS

AN IMPRINT OF SPRINGER-VERLAG

Published in the United States by Copernicus, an imprint of Springer-Verlag New York, Inc.

Copernicus
Springer-Verlag New York, Inc.
175 Fifth Avenue
New York, NY 10010

Library of Congress Cataloging-in-Publication Data

Harris, Sidney.
    Einstein atomized : more science cartoons / Sidney Harris.
        p.   cm.
    ISBN 0-387-94665-9 (soft : alk. paper)
    1. Science—Caricatures and cartoons.   2. American wit and humor,
Pictorial.   I. Title.
NC1429.H33315A4   1996
741.5′973—dc20   96-2616

Manufactured in the United States of America.
Printed on acid-free paper.

9 8 7 6 5 4 3 2 1

ISBN 0-387-94665-9   SPIN 10524006

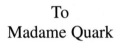

To
Madame Quark

"WE NOW KNOW ALL THE EXTRAORDINARY CHANGES THE UNIVERSE WENT THROUGH IN ITS FIRST SECOND. AFTER THAT, UNFORTUNATELY, IT TURNS OUT TO BE VERY MONOTONOUS."

"MY SECRET AMBITION IS TO FIND A NEW ELEMENT, BUT TO TELL THE TRUTH, I DON'T KNOW WHERE TO LOOK."

"DR. SYKES IS VERY IMPORTANT IN SCIENTIFIC CIRCLES. WE BELIEVE HE MAY BE THE MISSING LINK."

"ALL RIGHT—WHERE'S THE NUTRITION CRISIS?"

"IT'S NOT ENCROACHING CIVILIZATION THAT'S THREATENING OUR WAY OF LIFE — IT'S THE ENCROACHING HORDES OF PRIMATE-LIFE RESEARCHERS WHO ARE DOING IT."

"MORRISON IS LEAVING. HE BELIEVES HE'S FOUND ALL THE ANSWERS IN HERBS."

"UNDERSTANDING CREATION AND INFINITY IS WITHIN OUR GRASP... JUST AS SOON AS EVOLUTION PROVIDES US WITH LARGER BRAINS."

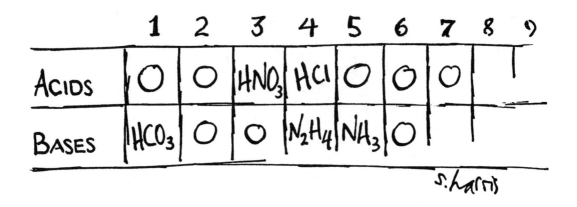

IF THERE WERE COMPUTERS
IN GALILEO'S TIME

"SOMETHING'S UP. THE INSECTS, THE TREES AND THE AQUATIC MAMMALS ARE _ALL_ PREPARING TO INHERIT THE EARTH."

TYPE-A SCIENTIST

THE AGE OF REPTILES

3 years, 2 months   One minute   114 years

AS THE UNIVERSE CONTINUED TO CONTRACT...

"THIS IS THE OLDEST THING WE'VE EVER FOUND. IT'S A FOSSIL OF A FOSSIL."

"YOU CALL THIS CLONING?"

"HE'S A PHONY — MADE OF ALL THOSE NEW
FAT SUBSTITUTES."

"I KNOW THIS IS ALL NONSENSE, BUT THAT'S THE PARADIGM I'M STUCK WITH."

COMPUTER-DRAWN, HAND LETTERED

HAND-DRAWN, COMPUTER-LETTERED

"WITH YOU, EVERYTHING IS GROSS SIMPLIFICATION."

"THE REASON HE'S NEVER SEEN A CONSTELLATION IS HE'S CONVINCED THERE <u>REALLY ARE</u> WHITE LINES CONNECTING THE STARS."

"IT'S NOT EASY TO HAVE A *SYMBIOTIC* RELATIONSHIP WHEN NO ONE ELSE IS AROUND."

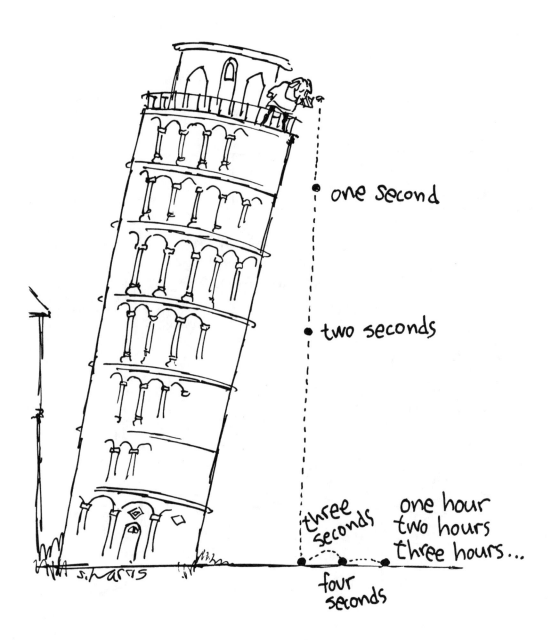

one second

two seconds

three
seconds

four
seconds

one hour
two hours
three hours...

s. harris

"EVERYWHERE YOU LOOK, THERE'S THE GREENHOUSE EFFECT.
IN HERE, WE CAN'T GROW A DAMN THING."

"IF EINSTEIN IS CORRECT, WHEN WE GET BACK, MY CAR WILL HAVE BEEN DOUBLE-PARKED FOR 320 YEARS."

"WHEN YOU GET OLDER, YOU'LL FIND YOUR TASTE WILL RUN TO BOAR, AARDVAARK AND BUFFALO... THE SLOWER ANIMALS."

TO AN OBSERVER APPROACHING
THE SPEED OF LIGHT, EINSTEIN AND
HIS SURROUNDINGS APPEAR TO BE
TALL AND THIN

"WHAT DID PEOPLE DO BEFORE THEY HAD ROBOTS?"

"I'M WORKING ON MY AUTOBIOGRAPHY, BUT LIKE ANY EXPERIMENT, I HAVE TO WRITE ABOUT THE GUINEA PIGS FIRST."

"OF COURSE I'M SURE IT'S A LEMON. IT CAN'T EVEN DETECT TERRESTRIAL LIFE."

"MY RESEARCH COVERS TWO FIELDS: THE BEHAVIOR OF MATTER UNDER HIGH PRESSURE, AND THE BEHAVIOR OF SCIENTISTS UNDER HIGH PRESSURE."

"THERE MUST BE SOME MISTAKE — HE'S LEAVING HIS MONEY TO SCIENCE, AND HIS BODY TO HIS FAMILY."

"TO THINK IT ALL STARTED WITH 'MELVIN HUNGRY, MELVIN WANT BANANA.'"

"WHAT'S COME OVER HEISENBERG? HE SEEMS TO BE CERTAIN ABOUT EVERYTHING THESE DAYS."

"IT'S FROM THE PROTOZOA RIGHTS COMMITTEE. THEY WANT TO KNOW WHAT YOU'RE USING THEIR CLIENTS FOR."

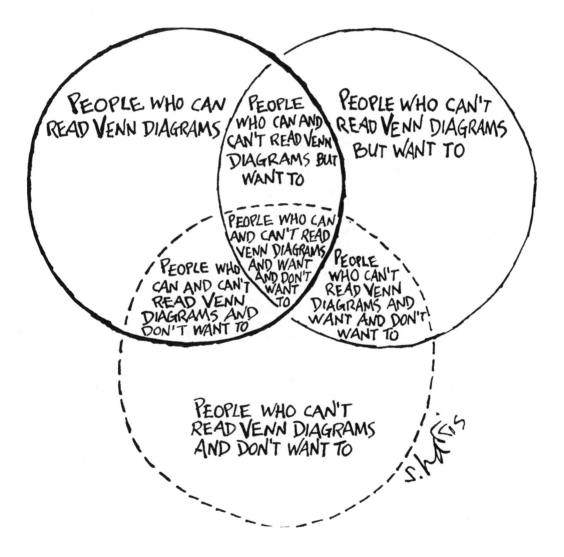

"ABOUT THAT X-RAY EMISSION FROM THE CONSTELLATION AR-215..."

LOGIC CHIP

INTUITION CHIP

BREAKTHROUGH BY A COMIC

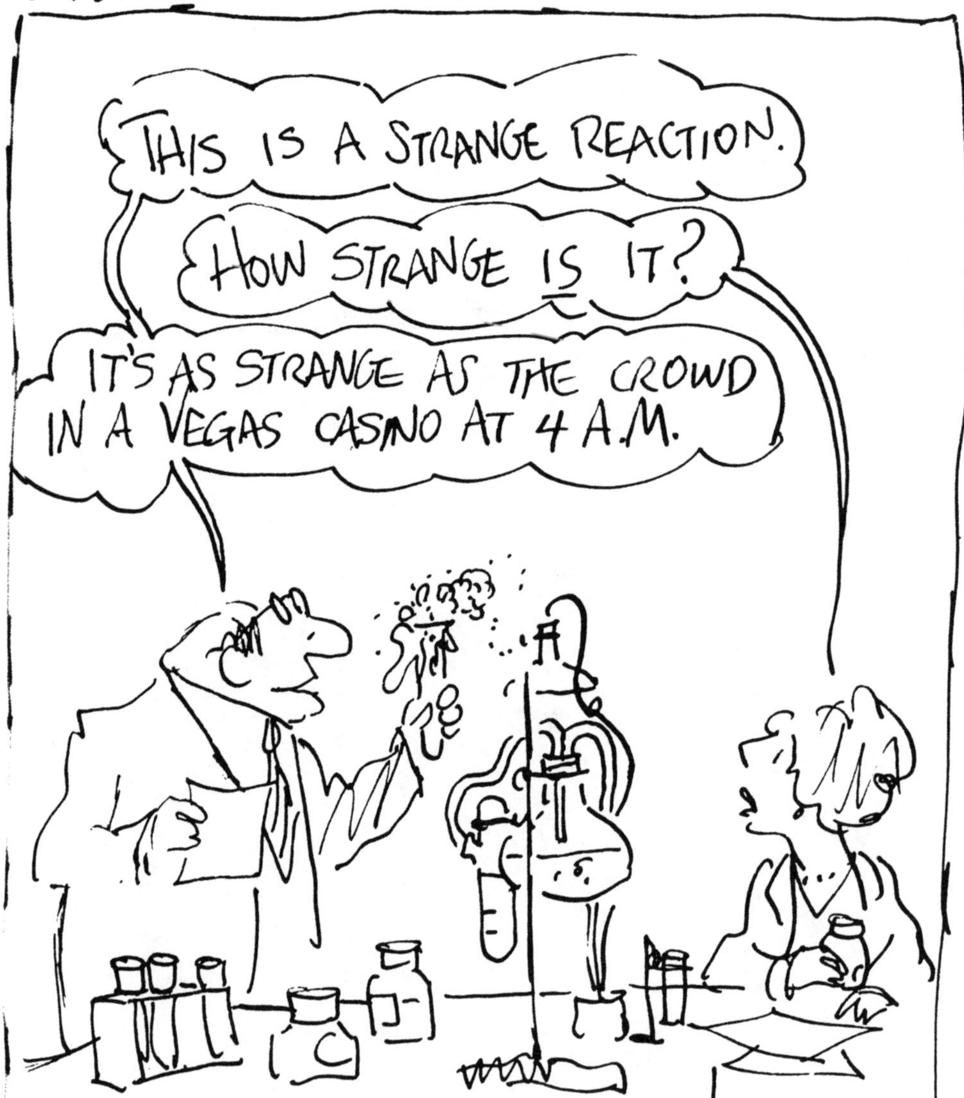

GALILEO'S BURDEN

BIG-HEARTED DR. QUARK
SHOWS THE DOG A BONE

"YOU CAN PUT AWAY YOUR TRANSLATING CALCULATOR NOW. I'M SPEAKING TO YOU IN ENGLISH."

THE BACK OF THE ROSETTA STONE

"THEN I SAY TO MYSELF, 'WHAT'S THE USE? THERE ISN'T ANY NOBEL PRIZE FOR MATH'."

FOR THE CHEMIST ON THE GO:
LAPTOP GAS CHROMATOGRAPHY/MASS SPECTROMETER

Krenwik wins Nobel Prizes

August Krenwik's great novel "Look Out, All My People!", which describes a remarkable device used to neutralize all explosive weapons, has led to his winning the Nobel Prizes for literature, chemistry, physics and peace. Because the book does not provide any information dealing with biology or health, he did not win the prize for medicine, nor was he in the running for the economics prize. "I am very disappointed I only won four of the prizes" he announced.

August Krenwick, born in Vienna, now lives in Memphis, Tenn.

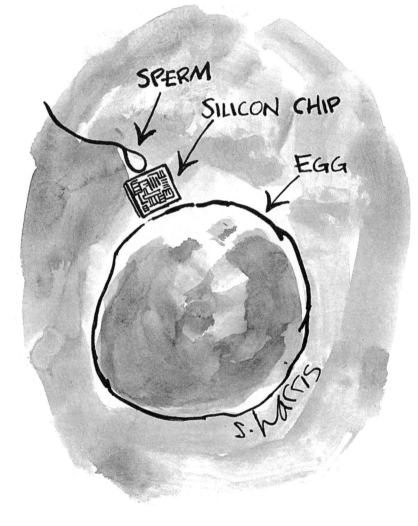

"TED, THIS IS Ms WARD. SHE'S COME TO US FROM 'ANALYTICAL AND FORMULATIONS CHEMISTS TEMPORARIES'."

"WE HAVE AN AGREEMENT WITH THE DEPT. OF AGRICULTURE TO ACCELERATE SOME GRAPES, LEMONS AND HAZEL NUTS."

"IN THIS FIELD, DUDLEY, YOU MISS A FEW HOURS AND
YOU'RE NO LONGER ON 'THE CUTTING EDGE.'"

MANY CREATURES HAVE ADAPTED BEAUTIFULLY TO THE HARSH LIFE OF THE DESERT
If I survive one more day here, it'll be a miracle.
I can't remember the last time I had any water.
This must be the worst place on earth.
It's like a furnace here.
S. Harris

"OF COURSE IT'S EFFICIENT. IT'S A
FLOW CHART."

"IF IT DOES COLLAPSE INTO ITSELF, IT WILL BE THE LAST WORD ON CONTEMPORARY PACKAGING: THE DISPOSABLE UNIVERSE."

TOPOLOGIST AT A PICNIC BRINGS THE LEMONADE IN A KLEIN BOTTLE

"THAT'S THE FUN OF BEING ORGANIC — IF ANY SALAD DRESSING IS LEFT OVER, I'LL USE IT AS SHAMPOO."

"WE WERE DOING A STUDY ON FABRICATING DATA. HE FABRICATED THE DATA."

I FEEL GREAT!
You should— you're getting younger.
BUT I SEEM TO BE FORGETTING EVERYTHING.
Get back in there,—quick!
WHY?
I BECAME YOUNGER? IF WE DO THIS FOR EVERYONE, WE'LL MAKE A FORTUNE.
You also forgot everything you knew. Do you want the world to be overrun by dopey kids?
IT ALREADY IS.
ALL RIGHT—I'LL CHANGE THE DEVICE INTO SOMETHING USEFUL — A MICROWAVE BATHTUB.
S. Harris

"AS LONG AS WE STICK TO STRONTIUM AND CELSIUM, WHICH REMAIN LETHAL FOR ONLY SEVEN HUNDRED YEARS, ALL OUR PROBLEMS ARE SOLVED."

"SHE'S BEEN WATCHING US FOR YEARS. WHEN THE HELL IS SHE GOING TO WRITE HER BOOK?"

"UP HERE, LIGHT IS NEITHER A PARTICLE OR A WAVE. IT'S A LIQUID."

"THIS YEAR, CHRISTMAS IS EASY FOR ME. I'M SENDING ALL MY GIFTS THROUGH MY MODEM."

"SOMETIMES IT DOES, SOMETIMES IT DOESN'T."

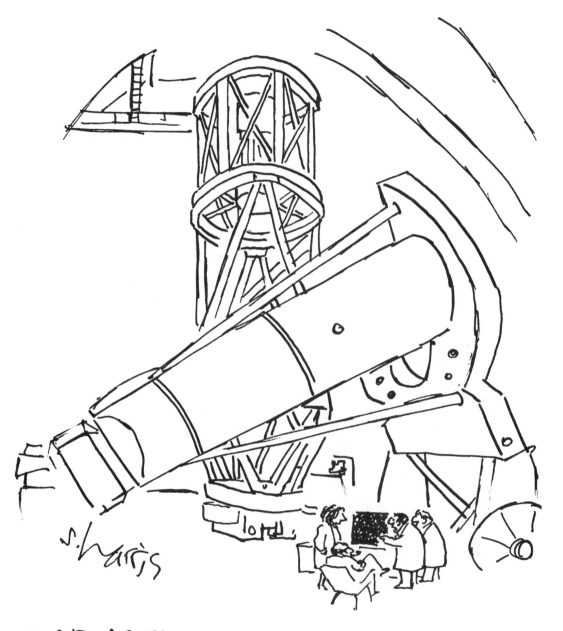

"...AND AT THE EDGE OF THE UNIVERSE IT SAYS 'ABANDON HOPE, ALL YE WHO LOOK BEYOND HERE'."

"ONE HUNDRED MILLION NEUTRINOS ARE PASSING THROUGH OUR BODIES EVERY SECOND, AND WE'RE WORRIED ABOUT THE PRICE OF COFFEE."

"AND YET, THE BUILDING IS CONCRETE, THE TANKS ARE METAL, THE WINDOWS ARE GLASS..."

"SURE, GENETIC FINGERPRINTING CAN PROVE SOMEONE'S INNOCENCE — BUT NOT IN A CASE OF STOCK FRAUD."

"I'M MAJORING IN COMPUTER SCIENCE AND I'M MINORING IN KINESTHETIC ANNUNCIATIVE AGGREGATES — YOU KNOW, TECHNOBABBLE."

EINSTEIN ATOMIZED

THE RESEARCH CONTINUES...

DO COSMOLOGISTS
FORM CLUSTERS?...

...AND DO THE CLUSTERS
FORM SUPERCLUSTERS?

j. harris

"THIS IS TERRIBLE. IF THE PLANKTON GO OUT ON STRIKE, IT'LL DISRUPT THE ENTIRE FOOD CHAIN."

"ON THE OTHER HAND, HE NEVER LEARNED TO RIDE A UNICYCLE."

"It's a wonderful find, and yet there's something suspicious about it."

"DR. GROMMET IS FUNDED BY A MAJOR HOLLYWOOD FILM STUDIO. HE'S BEEN ASKED TO COME UP WITH AN ANTI-GRAVITY DEVICE AND AN INVISIBLE-RAY GUN."

"THIS GALAXY SEEMS TO HAVE BEEN CREATED BEFORE THE BIG BANG, AND IT'S COMING TOWARD US. WE MAY HAVE TO RE-THINK SOME OF OUR OLD THEORIES."

"HOW'S THE NEW
INSOMNIA PILL GOING?"

PERPETRATOR OF A DARING, DAYLIGHT ILLEGAL COMPUTER TRANSFER OF FUNDS FLEEING THE SCENE OF THE CRIME

AT THE PRIMATE SPEECH CENTER
MOKOBO TRIES SOME STAND-UP COMEDY

FERMAT'S FIRST THEOREM

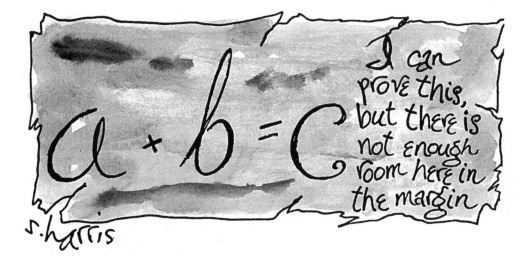

"SO YOU BOUGHT A FAX MACHINE! EVEN IF YOU HAD SOMETHING TO SEND, YOU HAVE NO ONE TO SEND IT TO."

"ACTUALLY I CAN'T EVEN FIGURE OUT THE ORIGIN OF THIS <u>ORGANIZATION</u>!"

NEVER
FORGETS

SOMETIMES
FORGETS

ALWAYS
FORGETS

CLOCKWISE

COUNTER-
CLOCKWISE

s. harris

"FOXCROFT, YOUR RESEARCH ON THE IMMUNE SYSTEM, AND THE DRUGS YOU'VE COME UP WITH ARE EXTRAORDINARY. AS A RESULT, WE'RE MAKING YOU A DISTRICT SALES MANAGER."

THE GRAPHOLOGIST AT WORK

Anton Chekov

Attila the Hun

Christoph. Columbus

S. Harris

Spiky letters... doesn't understand human nature. Small vowels...unimaginitive.

Should remain a physician, and give up foolish plan to become a writer.

Downward slant... timidity. Capitals... withdrawn, unaggressive. Should learn to be assertive.

Closed 'C'... desire to stay close to home. Wavy underline... fear of water. Would be happiest as a shepherd.